UP, UP AND AWAY

The Flight of Butterflies & Other Insects

UP, UP AND AWAY

The Flight of Butterflies & Other Insects

John
Brackenbury

First published 2022

Published under licence by Brown Dog Books and
The Self-Publishing Partnership Ltd, 10b Greenway Farm, Bath Rd, Wick, nr. Bath BS30 5RL

www.selfpublishingpartnership.co.uk

ISBN printed book: 978-1-83952-480-6
ISBN e-book: 978-1-83952-481-3

Cover and internal design by Lexi L'Esteve

Printed and bound in the UK

This book is printed on FSC certified paper

'What is this life if, full of care,

We have no time to stand and stare.'

WH Davies, 1871–1940

I am grateful to my friend and colleague Roy Barlow for his support and expert input over the many years of this project.

I. Introduction

Flight: the seemingly magical ability of animals to defy gravity and cleave the air as effortlessly as a cheetah glides across the ground or a fish slides through the water. I have been fascinated by the subject of animal locomotion ever since I became a zoology student many years ago and although many things which puzzled me then are much clearer in my mind now, the sense of wonder has in no way diminished. The author Douglas Adams summed up the feeling very well when he said 'I'd take the awe of understanding over the awe of ignorance any day'. In the modern world we are surrounded by a myriad everyday but highly sophisticated machines and gadgets: the fruits of two centuries of scientific and technological advancement. These 'black boxes' come in a huge variety of shapes and sizes, from cars to miniature hearing aids. We take them all for granted and even acquire expertise in the use of some of them; but asked exactly how they work and who invented them, we would be at a loss to give an answer.

Occasionally it is instructive to step away from our hi-tech world and spend a few minutes trying to cast our minds back to an age when ordinary people had much simpler lives, told the time of day by a sundial, understood that North lies in the direction of the Pole star, and knew every footpath in the vicinity because they trod them every day. Some individuals, spared for various reasons from humdrum chores and duties, turned their minds to the study of the natural world. These were the first 'free-thinkers' or philosophers, a privileged group but nonetheless respected by the community for their teachings and wisdom. Many of their ideas of how the physical world operates sound bizarre to us today but they did not have the advantage of a body of accumulated knowledge to call upon, at least nothing that was not already cloaked in myth, suspicion or fear. Scientific aids had not yet been invented: Galileo would not be born for another thousand years. The only tools at their disposal were their own eyes, imaginations and sense of curiosity.

I picture a small gathering of such enthusiasts upon a remote hilltop as evening begins to descend. In an atmosphere of calm expectation they await the appearance of the first twinkling treasures of the night sky. Even in the twilight of dusk, well before the sky has darkened, practised eyes already discern in the southern sky a line of bright dots. From previous observations they had worked out that these dots moved across the sky independently of the backcloth of the fixed constellations. These 'rogue' stars were actually the planets Venus, Jupiter, Mars and Saturn. One of the 'night-watchmen', his mind wandering a little, notices sinuous, pencil-thin skeins of geese disappearing towards the western horizon. Distracted for a while from their astronomical observations, the group exchange ideas on the possible nature of the power, divine or otherwise, that prevented the birds from falling to the earth. Was it the same power that kept the stars and the planets suspended in the heavens? Someone mentions that he had once been master of a sailing ship; he was familiar with the rules and conventions of navigation by the stars but also had a keen understanding of the power of the wind. Perhaps the wings of birds work like sails, he suggests. Heads respectively wag and nod but there is no consensus, the conversation begins to flag and they return to the more pressing matter of the planets.

Fast forward two or three thousand years and science has now taken the mysterious forces out of nature and replaced them with mechanical ones. The laws of physics and chemistry have withstood the test of time; wherever they have been applied they have been successful in explaining events in the natural world. A vast array of tools and instruments has been developed permitting scientists to probe deeper and deeper into the actual, as opposed to the assumed, nature of physical reality and new observations continue to surprise us and force us to review our interpretations. Who has not been surprised to learn about the sheer distances covered by birds, butterflies and locusts in their seasonal migrations across mountains, seas and deserts: all being brought to clearer light through the application of modern techniques such as radar-tracking and radio telemetry?

Human emotions run deeper than thought and our first response to nature is instinctively one of pleasure, awe or wonder. We raise a smile at the sight of a ribbon of butterflies chasing each other across the meadow and suddenly veering up into the sky, their twirling bodies climbing higher and higher until we can no longer see them. It is a gleeful sight, for a few enthralling moments we wish we could join them. We marvel, but alas only from the outside. We can never engage with a butterfly in the same way that we can with a pet cat or dog, or even a parrot. Yet there are occasions when the gulf can at least be narrowed, as we shall see later.

The human race will always see the sky as a metaphor for hope, freedom and the escape from worldly cares. At the time of writing, Britain along with the rest of the world is in the middle of a pandemic which has forced upon its citizens unprecedented restrictions on free movement and social contact. People feel the acute pain and discomfort of a forced closeted existence or having to maintain a prescribed 'social distance' whenever they meet someone else in the street. Perhaps never before have the words 'free as a bird' symbolised the yearnings and desires of so many people in so many places at the same time.

This book is my attempt to convey an impression of the world of flying insects, through the medium of photography. I hope the pictures will speak for themselves but some readers might also be interested in the story behind them: where the ideas came from and the processes through which the ideas were transformed into imagery.

To accompany the book, I have produced a collection of high-speed video recordings of insects in flight which readers can access on YouTube by typing in the search field: Up, up, and away: the flight of insects.

II. How Animals Fly

We know from the fossil record that a huge variety of animals has appeared upon the earth during the course of perhaps a billion years of evolution; yet only three types - birds, bats and insects - have truly settled for a life in the air. On the face of it this might seem odd because flight as a mode of transport offers obvious advantages: for example, the freedom to cover large distances at speed; to migrate seasonally from one place to another where living conditions might be more favourable; to avoid predators and physical encumbrances on the ground ; to escape fires and floods that engulf their earthbound relatives. Despite these and many other potential benefits, when you look up at the sky the aerial corridors by no means appear to be choked with animal traffic.

I think part of the answer to this question lies in the word 'specialisation'. Today's world is full of specialists, a condition forced upon us whether we like it or not, since every aspect of human endeavour has become so complex and highly competitive. Generalist have almost become an extinct species. A similar line of reasoning can be applied to the evolution of animals and plants, albeit on a much longer time-scale measurable in millions of years. Charles Darwin introduced us to the basic biological rule that organisms which adapt best to the prevailing environmental conditions are the ones that survive. The analogy with business and enterprise in human society is inescapable. But over-adaptation can drive an organism into an evolutionary cul-de-sac from which it cannot return if environmental conditions undergo rapid change. A brief 'cost-benefit' analysis of the evolution of animal flight will put this into context. We start by taking as given, that abandoning a life on the ground for one in the air is a very risky evolutionary adventure because it requires fundamental changes in almost every department of the animal's life: its behaviour, the structure and physiology of its body - not least its brain. To steer a course through the air, perhaps in buffeting winds, and then to execute a smooth landing, demands coordination skills of a new order. In human terms, it is the difference between learning to drive a car and training to become a test pilot. The biological template inherited from the flightless ancestor has to be radically modified and improved in order to manage these challenges. Not all insects have undergone this adaptation in the same way or with the same degree of success. Ladybirds and other beetle species, for example, are relatively weak flyers which puts them at the mercy of the wind; crash-landings are the norm but their armour-plated bodies are well-adapted to withstand injury. By contrast the flight of hoverflies, dragonflies, hunting flies and butterflies is far more accomplished, whether it be a brief sortie in pursuit of prey, a long day spent in energetic pursuit of females or a migratory journey crossing hundreds of miles of land or sea.

If indeed flying is such a perilous way of life, one could legitimately ask whether there are any examples of 'lapsed' flyers, animals that reverted back to the relative safety of the ground. It is often stated that evolution, like Time itself, cannot go backwards; but this is not wholly true. Wingless beetles and ostriches are just two examples of animals whose original ancestors were capable of flight but for some reason re-adapted to the ground: this phenomenon is known as 'secondary adaptation'. In contrast with these 'devolved' species of insect and bird, bats could be said from an evolutionary point of view to have become 'in so deep, there is no going back.' There are no examples of bats having secondarily adapted to a life on the ground for the simple reason that their hind legs have become irreversibly integrated into the wing: an extreme example of specialisation.

On numerous occasions I have watched the antics of butterflies finishing their feeding for the day and starting to search for a roost for the night. Prompted by its internal clock, the insect usually starts making preparations an hour or two before sunset; but if the air is warm and the sky is still bright, it might carry on feeding for an extra hour or so. Butterflies are very fastidious in their choice of overnight quarters; if a predator should strike in the night or early morning the fragile insect is defenceless. So as the light begins to fade, the behaviour of a late feeder takes on an extra sense of urgency. A Marbled White, for example, patrols restlessly up and down the same 50-metre stretch of grass at the edge of a wood, hovering here and there inspecting dozens of possible landing sites and becoming more and more agitated in its movements. Finally it selects a grass stem which to you or me looks indistinguishable from the thousands of others. Satisfied that the roost is safe, it shuffles through 180 degrees and settles head down for the night. Whether or not it sleeps, in the way that people sleep, I cannot say. But it certainly does not close its eyes. Insects have been highly inventive throughout their evolutionary history but the need to develop eyelids has never arisen. Each time human beings, mammals and birds blink, the eye is bathed in tears which moisten and lubricate the surface of the cornea. But the insect has a dry eye; in vain will you search for eyelids or any sign of a teardrop!

The large-scale navigational feats of butterflies have been well-documented: one thinks of the annual migration of the Monarch from Canada to Mexico or the familiar Painted Lady, which by the time it arrives on British shores may already have covered hundreds or even thousands of miles on its journey from Europe and North Africa. The geographical movements of insects can even be tracked using hi-tech instrumentation, including radar. But you can also learn a lot about the orientational skills of butterflies just by patiently using your own eyes. There is no better place to start than a wildflower meadow with tall grasses such as Timothy, Cocksfoot and Bromes, the favourite haunt of Meadow Brown, Marbled White and Ringlet butterflies. Their daily routine is divided between long, leisurely periods of feeding interspersed with short journeys through the meadow, during which they trace elaborate paths deep within the maze of grass stems. They always remind me of tiny helicopters moving stealthily through a dense forest of pine trunks. Even when the grass is swaying in the wind they never seem to brush the tips of their wings against the grass stems. This deft display of 'wing-work' takes place under the guidance of a brain containing only a million nerve cells or thereabouts. For comparison the human brain has almost a hundred billion (100,000,000,000) nerve cells.

Virtually every minute that they are in the air, flying insects are performing complex aerial manoeuvres (this does not apply of course to beetles). They are continually reacting to events around them: the sight of a predator, an item of prey coming into view, a potential mate, a rival male invading their territory, a favourite species of flower, an obstacle suddenly appearing in their path. Their reflexes are much faster than our own, due to the very simple fact that the nerve impulses of a small animal only have to travel over short distances. For a supreme display of reflexes in action, watch a dragonfly twisting and turning in the air in pursuit of its prey. Fortunately, dragonflies are also remarkably tolerant of the presence of human beings and if you stand watching one long enough, it is very likely to land on you and use you as a convenient lookout post for passing flies and other small insects. This familiarity gives you an opportunity to do a very simple experiment. Once it has settled on a favourite perch, which might be you yourself or the top of a fence post, you will notice that its head is constantly swivelling. Although the eyes cannot move independently of the head, the head itself pivots on the body through a hairspring mechanism more delicate than anything you would find in a Swiss watch. In fact whenever I see it, I am reminded of the bobble-headed dogs frequently seen on car dashboards. If you try to fool the insect by dropping a tiny stone through the air next to it, the dragonfly will target the decoy with a swift turn of its head but just as quickly ignores it. Moments later a small fly zooms past and in a trice the dragonfly is literally onto its tail.

A hunting insect like the dragonfly survives on its power and speed in the air and since they are amongst the biggest of all extant insects – hawkers have a wingspan up to 12 centimetres – one might have thought they had reached the limit of body size compatible with manoeuvrability. Yet there are giant fossil species with a wingspan of 70 centimetres, about the same as the wingspan of a kestrel. I suspect that these leviathans must have been quite cumbersome in the air and, like the albatross, would barely have been capable of taking off from the ground. They would not have been able to survive in today's environment because birds such as hawks would find them a very easy target. I find this a more persuasive argument than the alternative which has been advanced, that higher atmospheric oxygen levels in the Carboniferous era allowed bigger and heavier animals to evolve than is possible today.

The simple stone-dropping experiment gives an inkling of the power of the dragonfly's visual system to discriminate between objects of different shape and size. Interestingly, the opposite seems to be the case in butterflies. The male Silver-washed Fritillary butterfly is a dashing-looking insect with bright orange-red wings and his chief purpose in life is to chase after females, most of whom reject him anyway because they have already mated with another male. Experiments have shown that if a butterfly-sized red object is trailed through the air he is just as likely to pursue it as he would a genuine female. I have seen the same behaviour in the males of all the common white butterflies in the UK: Small-, Green-veined-, Orange-tip and Large White. They too are obsessed with a particular colour, in this case white. As he speeds along in his anxious quest, he is compelled to investigate anything in his path which is white and roughly butterfly-sized. A chance encounter with a large patch of mop-headed dandelion docks or ox-eye daisies releases a neurotic impulse which ends in frustration but he carries on undaunted.

It could be argued that of all the special senses that an animal possesses, the most indispensable is neither vision nor smell nor hearing, but balance. Deprived of a sense of balance an insect, a mouse or a human being would not be able to fulfil the first requirements of an independent existence which are to distinguish between 'up', 'down' and 'side', to stand erect and to walk in a straight line. Many of us have had experience of how easily this primal sense can be impaired by a small intake of alcohol. We humans maintain balance and posture with the help a self-levelling device lodged in a bone in the inner ear. An infection in one or both ears has a similar effect to being drunk or seasick and may cause you to lose balance and become violently sick. Even an earthworm, which from a superficial glance looks perfectly cylindrical, has a flatter underside on which it slithers along the ground. In the case of a flying animal it is hard to overstress the importance of balance and once again the agile dragonfly provides a good example. I have high-speed video recordings showing the insect rolling its body through almost 90 degrees, like a fighter aircraft banking hard to the side, yet the head remains perfectly horizontal. For a brief moment the dragonfly looks as though it has literally 'twisted its neck'. I referred earlier to extreme mobility of the dragonfly's head on its body, and this provides a perfect gyroscopic mechanism for keeping the eyes level with the ground during flight manoeuvres.

High-speed videography can add a new dimension to understanding how insects maintain balance in the air. For example, it is surprising to see how often insects bump into nearby pieces of vegetation during take-off. These minor collisions are invisible to the naked eye because the insect makes such rapid adjustments. In fact the adjustments are so fast – virtually instantaneous – that they cannot possibly be due to active nervous control. There must be an automatic self-righting process at work. Bees and hoverflies seem especially accident-prone during take-off and the videos show them 'rolling with the punches'. A buoy provides a useful analogy. It is impossible to capsize a buoy because it is bottom-loaded and even if you did manage to push it right over it would immediately bob up on the other side. The insect's body is slung below the wings and this makes it bottom-loaded just like the buoy. The high-speed video camera is continuing to reveal surprises, including insects which for some as yet unknown reason launch into flight whilst doing a backward somersault. The project is very much in its infancy and no doubt even stranger findings will come to light.

III. Animal versus Aircraft

It is interesting to compare the guidance system of an insect with that of an aircraft controlled by a human pilot. When the plane begins its descent to the airport towards the end of the journey, the captain switches from autopilot to manual control. His hands are now on the joystick, his eyes now take over the job of monitoring the rapidly changing geometry of the terrain below. Anyone who has seen footage of a plane being tossed about by the wind as it approaches the runway will appreciate the problem the pilot is up against. In principle this is no different from a bee or a butterfly attempting to land on a flower flapping wildly in the breeze. Often the insect fails on the first few attempts, hovers uncertainly by the lip of the flower and only after a period of 'toe-dipping' finally makes full contact. In this situation the insect's brain is working hard to cope with a stream of information from its eyes, its organs of balance and the touch receptors in its feet. Distractions from nearby vegetation fluttering in the wind add to the problem. The same considerations apply to a bird coming in to land on a swaying branch. In this case, it is highly unlikely that the nervous system of the reptilian ancestor would have been able to process information at the necessary speed and one imagines that at some point during the transition from ground to air a quantum leap must have taken place in the sophistication of the wiring of the brain.

Hence safety is just as paramount in the design of a flying animal as it is in that of a passenger aircraft (I use the term 'design' in the Darwinian sense of living systems being acted upon and improved by natural selection). In both cases, if and when an accident happens in the air the consequences are likely to be catastrophic. If a fox sprains a limb joint or even fractures a bone in its leg, this does not automatically spell the end of the fox's world. But a broken wing in a bird or a butterfly is tantamount to a death sentence because the animal immediately becomes a target for predators on the ground.
If you think this is an exaggeration, consider the following: an investigation was once carried out to find evidence of old repairs in the skeletons of animals that had been dug up from various locations in the wild. An old repair in a broken bone is evidence that despite the injury, the animal was able to survive at least

long enough for the repair to take place. That could take months or years. The findings were good news for animals with four legs but disappointing for bipeds with wings. Plenty of old repairs were found in the leg bones of the mammals but not a single case in the wing bones of a bird. The conclusion has to be that in the wild a bird rarely if ever survives long enough to repair a fractured wing bone. And the same surely goes for a butterfly with a broken wing. Towards the end of the butterfly season it is very common to see individuals still managing to fly on very tattered wings, sometimes having lost large areas of the wing to bird strike. But a clean break in the leading edge spar of the wing - the equivalent of the humerus, radius and ulna bones which brace the wing of a bird - will invariably prove to be fatal.

Any study of the mechanics of animal flight inevitably begins with a comparison with man-made aircraft and there are many points of convergence .Birds have a single pair of wings just like an aircraft and this separates them both from insects which have two pairs, the only exception being the true flies or Diptera. When the pioneers of human flight took to their workshops they instinctively based their designs on bird wings not insect wings, because bird wings are the equivalent of human arms. But the model was fatally flawed from the beginning, both metaphorically and in practice since many of the daring 'ornithopter' pilots actually did plummet to their deaths. We look back in disbelief at vintage footage of men attaching artificial wings to their arms, then leaping off cliffs and vainly attempting to replicate the flapping motion of a bird. In so doing they vastly underestimated the amount of power needed to suspend a body the size and weight of a human being in the air. Had their knowledge of physiology and biomechanics been more advanced they would have seen that the pectoral muscles of a bird are enormous, accounting for as much as 50 per cent of its entire body weight. Human pectorals are by comparison puny. If a person of say 70 kilogrammes body weight were ever to get off the ground using wing power alone he or she would need approximately 30-40 kilogrammes of that weight to be transposed to his or her breastbone in the form of sheer muscle, and the breastbone itself would need to reach down to his or her knees. People were never made to fly like birds! Looking back, one can see how these failed experiments were perhaps necessary in order to pave the way to more successful prototype designs based on a seated fuselage with fixed wings and powered by an engine: the true predecessors of modern aircraft.

Animal wings work quite differently from aircraft wings. Put at its most basic, the job of an aircraft wing is to produce the lifting force which keeps the plane up in the air, but it can only do this if the aircraft is already being driven forward by its engine. Call it a division of labour. But animal wings have to do both jobs at the same time: create the forward driving force as well as the lift. That is why they flap. It is so obvious you would hardly think it worth mentioning but if you constructed a model aircraft the size of say a crow, equipped it with regular aircraft-type wings which could also flap up and down, the machine would tumble straight to the ground. A totally rigid flapping wing cannot work. To envisage animal wings as the equivalent of cat flaps or trapdoors hinging up and down on the body, would be a complete misconception. Real wings are flexible, they continually bend and twist throughout the stroke in a highly regulated manner. The left and right members of a wing pair can each produce lift independently, just as they do in an aeroplane, but in addition they can interact with one another to produce extra lift. This is not possible if the wings are fixed and rigid. You can actually hear the effects of wing interaction when a wood pigeon bursts noisily out of a tree. The loud clapping sound is due to the wings hitting against each other at the top of each stroke. Then, as the two wings separate to begin the downstroke, they produce a suction force above the body. This description is based on birds but butterflies and many other insects have their own version of the so-called 'clap-and-fling' mechanism and I suspect that if our ears were sensitive enough we would hear a round of soft 'applause' each time a butterfly fluttered past. The uncovering of the unique characteristics of animal flight is an example of what is now referred to as 'interdisciplinary science'. Biologists, mathematicians and physicists together set out to solve a conundrum which had been for decades, and in some people's minds still is, a piece of modern folklore: according to theory bumblebees cannot fly. The problem of course was not the bee but the theory: bees are not aircraft and have worked out their own ways of cheating gravity.

Leaping bush-cricket.

Spider-hunting wasp Sceliphron.

Bush-cricket Conocephalus.

Leaping Mantis nymph Empusa.

Mantis Iris oratoria.

Ichneumon fly.

Leaping Mantis Ameles.

Flea beetle.

Wasp experiencing a glancing blow from a droplet.

Mantis.

Cicada. Note the propeller-like twist in the wings

Long-horn beetle.

Beetle *Metoecus paradoxus*.

Red-underwing moth.

Jewel beetle Julodes.

Soldier beetle Cantharis.

Scorpion fly Panorpa.

Skipper Ochlodes.

Digger wasp Sphex.

Great Grey Bush-cricket Decticus.

IV. A Photographer's Insights into the World of Insects

My interest in insect flight was triggered purely by scientific curiosity. When you examine the wings of a dead insect caught in a spider's web or pinned to a board in a display cabinet, they seem anything but, as the saying goes, fit for purpose. Dry, brittle and kinked in places they would probably snap at the very first stroke in real life. In a living insect the fabric of the wing is kept hydrated, supple and elastic because it receives a continuous blood supply via the wing veins. The wings need to be flexible in order to work at all. In addition, adjacent sections of the wing membrane can move relative to one another along strategically located fold lines, rather like miniature tectonic plates. To a certain extent it is possible to speculate how these various movements might affect the wing profile during flight, just by manipulating fresh specimens. But the only way to be certain is to carry out detailed investigations on the freely flying insect. This was the context in which I set out to do my study. The first task was to build the equipment: a high-speed photographic system which would be fast enough to freeze the motion of the wings of free-flying insects and sharp enough to show their slightest distortion, right down to the tiniest folds and wing veins. Capturing the insect in flight would be achieved through a remote-controlled triggering device, easy to say but difficult to achieve in practice. In order to design and assemble the equipment I would need the help of someone trained in electronics. Engineers who also happen to be interested in insect flight are a rare breed but to my great surprise I discovered the ideal individual working in the electronics section of my own university department. The collaboration that followed proved so productive that soon we were able to publish a joint technical paper giving details of the circuit design. That was nearly thirty years ago. Today I can look back on this piece of good fortune and realise how significant it proved to be in everything that was to follow. It made me appreciate that when you take on a very challenging exercise, a point will arrive where you need help from others who are far more skilled in certain areas than you are. Whatever you go on to achieve may well be in large part due to their selfless willingness to engage at least for a while in the pursuit of your dream. The outcome of this particular partnership was that after two years I was able to introduce to the world of insects my prototype high-speed photographic system.

Over the following years I took thousands of images – many of them during visits to northern Spain with its rich fauna of insects – depicting every conceivable twist and turn in the wings of almost every class of insect that could fly. I have included some of the photographs from that period in the first gallery of this book. But I also have to say that I was relieved when the project was over because it had suffered a succession of mechanical problems, none of them actually insoluble but bad enough to eat away at one's enthusiasm. So at that point I was happy to say goodbye to the rather obscure subject of insect biomechanics. But events took a turn and I was soon back on the trail with a different project.

I am wary of that well-known saying 'what does not kill you makes you stronger' which sounds artificial and glib and often turns out to be catastrophically wrong. But on this occasion I think there was some truth in it. After you have devoted several years of your working life, day in, day out, to making observations on an animal – even as part of a rather dry, clinical investigation of the mechanics of an insect – you begin to see the animal in a different light from when you started. Initially it is no more than a collection of data points for a scientific study, a biological machine on which you make elaborate measurements. You eventually realise that the 'machine' is in fact a complex living thing, with a life outside your immediate concerns. Their relationships with the wider world may seem to be of an infinitely lower order if judged by human standards but that is not the correct way to view them. Perhaps no one understands this better than beekeepers, who come as close as is humanly possible to interpreting the mind of an insect. It is also true that most insects are just as capable of observing you as you are of them: there are two ends to a telescope. I wouldn't want to overstate this, insects are an alien species to which a human being represents only one of two things: either a large threatening presence or else a desirable target for a meal of blood. Enthusiastic owners of pet mantises and tarantulas might take a different view but I think it applies equally to a butterfly evading capture from someone brandishing a net, or a mosquito intent on piercing your arm in the middle of the night.

This does not necessarily rule out the possibility of a more perhaps 'nuanced' interaction between man and invertebrate. The question may not be as strange as it sounds because almost all of us have at some time been in a situation where the notion could be said to apply. You are cooking, the kitchen is full of steam and appetising smells so you open the door to the outside. Suddenly you hear the drone of a bluebottle. You wave your hands vaguely as it tries to settle near the cooker. The fly keeps coming back, even has the nerve to land on your arm; so you seize a spatula, swipe wildly at the air and chase the hated intruder from the kitchen to the sitting room. No sooner have you returned to the cooker than the drone starts up again. For a moment the pitch of the sound rises as the bluebottle flies up and down the walls and the window panes, anywhere so it seems but the open door. 'Stupid fly!' you mutter, bearing down on it once more. The fly now senses the increased menace in your movements, suddenly remembers where the door is and flies straight out. 'Not so stupid after all', you think. It is difficult to avoid the impression that during the encounter the fly has been weighing the risk of possible harm to itself against possible reward in the form of food. The military concept of 'graded' or 'tactical' response to threat would not be a million miles away from an accurate description of the fly's behaviour, even though the word 'cantankerous' sums it up more succinctly.

For some reason I am reminded at this point of the curious 'observer effect' identified in an area of science that could hardly be more remote from insect behaviour: quantum physics. We are told by specialists in this field that experimental observation can never truly reveal the ultimate reality of matter because once you get down to the subatomic level, particles themselves are altered by the act of being observed. This is a very strange and in some ways unsettling finding but it has also had the beneficial effect of bringing together physicists and philosophers. I understand only enough of the theory to mention it in passing but I cannot help wondering whether the observer effect might also be of interest to animal behaviourists, even those engaged in the study of 'lowly' animals such as insects.

Anyone who has ever tried to photograph a butterfly feeding on a flower, let alone one flying through the air, will be familiar with the frustrations of the exercise. You creep closer and closer and just at the moment when you are about to press the shutter button, the insect flies off. You are allowed to get close, but not quite close enough. You try again, then again and one by one the next ten butterflies thwart you in exactly the same way. With patience running out you decide to have one more try and inexplicably the eleventh subject cooperates. The eleventh is no less alert than the other ten were; they all show the same behaviour pattern, periodically pausing in their nectar-feeding, withdrawing the proboscis and lifting their head as though intently watching you. Eventually you learn that when the insect pauses, that is the moment for you too to pause; when it resumes feeding, that is your opportunity to steal a little closer. And so on. It is a game of bluff and you quickly learn to read the signs, understand where the boundaries lie. As does the butterfly.

After thousands of episodes like this you come to the conclusion that butterflies are like people in the sense that no two are exactly alike. Also, simply by being so physically close to the insect makes you aware of how much more calculated its responses are than you might have expected. Inevitably in a situation like this, the question of 'intelligence' arises. Unfortunately the word itself is a minefield and in the human realm has probably done more harm than good. As a sociological concept it is divisive, it polarises debate and it gives self-appointed experts the licence to pass quasi-scientific judgements on others. I think it is presumptuous to say of any human being that they have been shown scientifically to have an IQ of this or that, as though their abilities were measurable in the same way as their blood group. An IQ test cannot be objective because no one would design such a test without choosing criteria that guaranteed them personally a place in the upper quartile of the distribution curve. They would be silly not to. Perhaps the test should be renamed the AT or 'aptitude test' in the sense implied by: 'The eminent psychologist Dr Higgins, author of many acclaimed IQ tests, has himself sat them many times and finds that he has a great aptitude for them'. One of the finest brains by any measure was Albert Einstein, who is reputed to have said, 'Everybody is a genius. But if you judge a fish by its ability to climb a tree, it will live its whole life believing that it is stupid'. Incidentally, I have never heard a better definition of the intelligence of fish. Of his own achievements the great man remarked that it wasn't so much that he was smarter than everyone else but he could stick at a problem longer than everyone else. A little modest but nevertheless a very generous and inclusive answer which gives hope and reassurance to us all.

V. A Novel View of Insects in Flight

As I outlined at the beginning of the last chapter, when the scientific project on which I had been working finally came to a close, I felt a sense of relief. I was also satisfied that many of the scientific questions I had originally posed had been settled by the results. As is the tradition in science, the findings were written up and eventually published in a specialist journal. For a while I moved away from the subject but I couldn't entirely detach myself. Something kept passing through my mind, not a new idea along the old lines, more like a series of fleeting images. Impressions are not the same as meanings and it is not easy to give shape and direction to them but they did eventually distil into something tangible enough to recognise and even give it a provisional name: an 'insect panorama', an image that reproduces the world view of an insect.

A brief foray into the world of close-up photography will help to clarify where the stream of thoughts was coming from. The conventional way of depicting very small subjects like insects is through the use of a close-up lens. It is called macrophotography and in this mode the camera works like a microscope, magnifying the subject and rendering it in crystalline, frame-filling detail. At the same time however the picture loses a sense of depth; the three-dimensional space around the central subject shrinks to the point where it becomes enclosed in its own spatial bubble, disconnected from the wider world which it properly inhabits – which is our world too. The effect is a natural consequence of the laws of optics and is in fact seen by macro-photographers as highly desirable because it eliminates distracting background noise. A high-street photographer looks for the same effect by posing his studio subjects against a blank background. The photographs in the first gallery were taken using this standard technique. People base their views of the world very largely on what they see in the visual media so the impression is easily given that insects are locked in a closed world of their own, with mental horizons extending no further than the tips of their antennae. How could such a purblind organism possibly share any part of the wider sensory experience of human beings? So the practical question for me became: can an insect see the world as we do as a rich tapestry of landscapes with trees, fields, far-flung hills and mountains, the sky, the sun and the clouds? This is patently not the case with the many species which spend their lives hidden from view in the undergrowth and the leaf litter; theirs indeed must be a myopic world. However, an insect which spends most of its life flying about in the air inevitably operates in a much wider arena of visual experience. We know for example that bees construct mental maps of their foraging territories extending a mile or more beyond the hive or the nest. Highly territorial butterflies such as the Painted Lady and the Red Admiral can spot intruding rivals at a distance of 50 metres or more. Monarch butterflies cross a quarter of the globe during migration and it is difficult to see how they could achieve this without some degree of distance vision. Even the experimentally demonstrated ability of bees to count up to three or more is a direct consequence of the ability to recognise spatial patterns of flowers.

I cannot actually prove that an insect can see the landscape around it like we do even though I can raise arguments in favour of the possibility. And I am sure that even if it could, it would not recognise features of the landscape like we do or attribute meaning to them in the same way. But my intention was never to establish a scientific truth as such; rather it was to give my own imagination a helping hand in its efforts to frame a novel type of image. The route might seem tortuous but the mind doesn't always work in straight lines. By recreating the visual experience of the insect through a camera lens, I would in fact be giving myself and others an alternative view of nature. Incorporating the landscape is already a radical departure from the conventional idea of close-up photography. We already know that the domed compound eye of an insect is capable of extreme close focus, in fact right up to the surface of the eye itself and this also needs to be reflected in the image. It seemed appropriate to refer to this kind of image as a 'panoramic close-up'.

To bring these various elements together, I now invite the reader to join me in a thought experiment. Imagine that by some piece of wizardry you have been transformed into a miniature human being, an inch-high Tom Thumb. Someone air-drops you into a Lilliputian world of mosses, ferns and flowers, and butterflies sailing above your head like eagles. You are now asked to describe what you see. Tom Thumb's eye becomes the metaphor for the panoramic close-up image. Through his eyes you will see the world as though you yourself were a butterfly watching other butterflies flying close by. So much for the mental 'software' behind the image. But what about the physical 'hardware' needed to turn the image into an actual photograph? By good fortune it turns out that from a technical point of view there is already an equivalent of Tom's eye in the world of optics and photography: a fish-eye type lens with an extremely small focal length, much smaller than that in normal use in still photography. It was now a matter of finding the lens with the right specification, fitting it to a suitable camera body and experimenting.

VI. Butterflies in Flight: the Wider View

The wide-angle lens would allow me to represent the insect, not as an isolated entity at the objective end of a microscope but as a figure in the landscape. The appropriate analogy in human photography would be the difference between a head-and-shoulders portrait of someone in the studio and a wide-angle shot of the same person outdoors, say in a field of yellow flowers with blue sky and white clouds in the background. Countless varieties of camera exist on the shelves of specialist shops and stores but none had been designed with this kind of image in mind; not in the least surprising, since there is no market for it. Consequently, I set to work designing my own system. I will spare the reader details of the years of trial and error that then followed but I finally arrived at a piece of hardware appropriate for the task.

At this point, with the camera snugly in my hand, I began to have my first serious doubts about my ability to achieve the goals I had set myself. Somebody once said, 'if you will the means then you must also will the ends'. All I really had to show for my efforts was a gadget in my hand and a theoretical picture in my mind and neither of these is of any use to a nature photographer if you lack the necessary fieldcraft. The practical challenge for me, was to get within a few centimetres of a flying insect and then to photograph it. A few preliminary attempts was enough to convince me that it was virtually impossible with a single shot of the camera, the insect is just too quick. Then I discovered when playing with the controls of the camera that it had a setting which allowed you to reel off twenty full-resolution images in a single high-speed burst. This was potentially a 'game changer'. Even so, the burst would last less than a second and would my reflexes be fast enough?

That question was answered as soon as I set foot in the field. The speed of events as you try to react to an insect flying past your ear or rising quickly from a flower, rules out any possibility of composing the picture either through the viewfinder or in the rear screen of the camera. Normally when you take a photograph you expect to see the subject in front of you. That was out of the question, so the only way I could think of to overcome the problem was to teach myself to shoot 'blind'. After innumerable episodes of mainly hapless flailing of the arms, I eventually reached a point where I could roughly guess the scene in the frame of the camera as I swung it into action. On the other hand, once I had pressed the shutter button it was largely a matter of chance whether or not I had captured the insect itself.

I had played squash for many years and I remembered a phrase which seemed relevant to what I was trying to do: 'court awareness', which in practical terms means making use of your 'peripheral vision'. Re-educating my eye and body to these very unfamiliar shooting conditions was like learning to play squash; developing an awareness of, and reacting to, movements in the corner of your eye. Possibly those earlier experiences, and the muscle memory that arose out of them, helped in these novel circumstances. People generally assume that insect photography is a leisurely pursuit but anyone spotting me brandishing my camera in the field might arrive at a very different conclusion. Nearly all my shots are taken from ground level, which I suppose is the closest I can get to being Tom Thumb himself. I shoot 'on the hoof', opting to go in pursuit of the quarry rather than waiting for it to come to me. A typical day in the field would find me up and down on my knees hundreds of times, often in quick succession and inevitably the physical and mental effort takes its toll. On some occasions frustration reached a point where I could easily have flung the camera at the nearest tree trunk. This might read like a litany of woes, but I doubt if it could have been any other way, regardless of who was wielding the camera.

At the end of a day of 3, 4 or 5 hours of photography in the field, I sometimes come home feeling tired and deflated, like someone who has found himself at the wrong end of a gruelling poker session. Your instinct is to blame everything: yourself, the insects, the weather – especially the infernal clouds, which are in the habit of marching across the sun just at the moment you need perfect light. Feeling slightly more philosophical after a cup of tea, you take out the camera and peruse the shots that have not yet been erased, the ones that passed initial scrutiny at the time you were shooting. Once again you review your methods, try to fathom out why they worked sometimes but not at others. Factors such as the weather – those clouds! – are beyond your control but can you improve your approach to the insects? Is it just a matter of timing (if only time could dilate, just as Einstein said) or are the butterflies still outsmarting you in some way? Perhaps 'outwilled' would be a better way of putting it. All animals express 'will' in the sense that they make things happen in the world, in contrast to an inanimate rock which cannot will anything. 'Will' is a neutral term, it just implies intention and doesn't carry any of the baggage associated with the word 'intelligence'. An amoeba can will itself to move but no one would claim it is intelligent. When I enter the open fields and woods, I step out of my own natural environment and into the insect's. The insect therefore has the advantage. And when I try to photograph it going about its business, I am pitting my will against its and the odds are heavily in its favour. Every wildlife photographer eventually learns lessons like this but I feel insect photographers have the added burden of finding themselves eye-to-eye with the adversary; this makes the encounter quite different from photographing a bird through a telephoto lens; and incidentally might also explain why nature photographers often find it difficult to switch from one discipline to the other.

Three years were spent in furious pursuit of these entomological will-o'-the-wisps and although I cannot say that the process got any easier, I did register more successes than I imagined possible and the pick of these is shown in the second gallery of images. I am only too aware that pictures are a poor substitute for the real thing but if they can encourage others to go out and observe nature through their own eyes, they will have served the purpose for which they were intended. I said 'through their own eyes' without thinking but did I really mean 'through the eyes of Tom Thumb'?!

As an accompaniment to the book, readers will find a compilation of high-speed video recordings of flying insects on YouTube. Type John Brackenbury, Up, Up and Away into the search field.

Large White butterfly Pieris brassicae.

Peacock butterfly over springtime Hawthorn blossom.

Brimstone butterfly Gonepteryx rhamni.

Peacock butterfly Inachis io.

Marbled White butterfly Melanargia galathea.

Meadow Brown butterfly Maniola jurtina.

(left and right) Reconstructing a flight sequence. A high-speed burst of the camera produces twenty exposures and increases the chances of capturing at least one good image of the insect in flight. In some cases the burst captures several consecutive images and these can be assembled into a single image representing the sequence. The panel includes four images from such a sequence in a Large White butterfly Pieris brassicae. All the images are virtually identical, except for the position of the insect. The first image in the sequence is used as the 'base' and successive stages of flight are merged with the base using simple computer software.

Brimstone butterfly Gonepteryx rhamni. I often try to aim the camera in the direction of the sun in the hope of getting dramatic lighting effects on the wings, and the gallery includes many examples of this technique.

Ringlet butterfly Aphantopus hyperantus. The butterfly has taken off backwards and for a brief moment it is upside down.

(left and right) Meadow Brown butterfly Maniola jurtina. These are both from the same sequence.

Reconstructed take-off sequence in a Green-veined White butterfly Artogeia napi.

Brimstone butterfly Gonepteryx rhamni.

Painted Lady butterfly Cynthia cardui.

Large White butterfly Pieris brassicae.

Reconstructed take-off sequence in a Red Admiral butterfly Vanessa atalanta.

Silver-washed Fritillary butterfly Argynnis paphia.

Large White butterfly Piersis brassicae.

Reconstructed flight sequence in a Wasp *Vespula vulgaris*.

Reconstructed take-off sequence in a Meadow Brown butterfly *Maniola jurtina*.

Brimstone butterfly Gonepteryx rhamni.

(left and right) Painted Lady butterfly Cynthia cardui. This and opposite are from the same sequence.

Reconstructed flight sequence in a crane fly. Note that in some frames the leading edge of the wings is facing backwards: this is because the wings have rotated through almost 180 degrees during the upstroke.

Marbled White butterfly *Melanargia galathea*.

Large White butterfly Pieris brassicae.

Red Admiral butterfly *Vanessa atalanta*.

Reconstructed take-off sequence in a Brimstone butterfly *Gonepteryx rhamni*.

Large White butterfly Pieris brassicae.

Dragonfly.

Dragonfly.

Reconstructed take-off sequence in an Orange-tip butterfly Anthocaris cardamines.

Chaser dragonfly copulation wheel.

Damselfly with prey.

Damselfly copulation wheel.

Reconstructed take-off sequence in a Painted Lady butterfly Cynthia cardui.

Meadow Brown butterfly *Maniola jurtina*

Reconstructed take-off sequence in a Small White butterfly Artogeia rapae.

Mating Green-veined White butterflies *Artogeia napi*.

Reconstructed take-off sequence in a Painted Lady butterfly *Cynthia cardui*.

Marbled White butterfly Melanargia galathea.

Peacock butterfly Inachis io.

Mating pair of Small White butterflies Artogeia rapae.

Reconstructed take-off sequence in a Marbled White butterfly *Melanargia galathea*.

Red Admiral butterfly Vanessa atalanta. This photograph is from the same sequence as shown on page 52.

Meadow Brown butterfly Maniola jurtina.

Brimstone butterfly *Gonepteryx rhamni*

Reconstructed take-off sequence in mating Chaser dragonflies.

Peacock butterfly Inachis io.

Reconstructed take-off sequence in a Brimstone butterfly Gonepteryx rhamni.

Red Admiral butterfly *Vanessa atalanta*.

Reconstructed take-off sequence in a Brimstone butterfly Gonepteryx rhamni.

Mating Green-veined White butterflies Artogeia napi.

(left and right) Mating Common Blue butterflies Polyommatus Icarus.

Reconstructed take-off sequence in a Large White butterfly *Pieris brassicae*.

Red Admiral butterfly *Vanessa atalanta*.

Reconstructed take-off sequence in a Chaser dragonfly.

Painted Lady butterfly *Cynthia cardui*.

Reconstructed take-off sequence in a Peacock butterfly Inachis io.

Brimstone butterfly *Gonepteryx rhamni*.

(left and right) Painted Lady butterfly *Cynthia cardui*.

Small White butterfly Artogeia rapae. This and opposite show the wings above the body in the 'clap' position.

Brimstone butterfly Gonepteryx rhamni.

(left and right) Common Blue butterfly Polyommatus Icarus.

Meadow Brown butterfly *Maniola jurtina*.

Reconstructed take-off sequence in a Small White butterfly Artogeia rapae.

(left and right) Large White butterfly Pieris brassicae.

Reconstructed take-off sequence in a mating pair of Green-veined White butterflies Artogeia napi.

Courtship behaviour in a pair of Green-veined White butterflies Artogeia napi. The upturned abdomen of the female signals to the circling male that she is already mated and rejects him.

Red Admiral butterfly *Vanessa atalanta*.

Small White butterfly Artogeia napae. The butterfly has taken off backwards and for a brief moment is almost upside down.

Reconstructed take-off sequence in a Large White butterfly Pieris brassicae.

Small White butterfly Artogeia rapae.

Green-veined White butterfly Artogeia napi.

Brimstone butterfly *Gonepteryx rhamni*.

Mating Green-veined White butterflies Artogeia napi.

Take-off sequence in a Ladybird reconstructed from two flights during an Autumn swarm.

Red Admiral butterfly *Vanessa atalanta*.

Marbled White butterfly Melanargia galathea.

Brimstone butterfly Gonepteryx rhamni.

Reconstructed take-off sequence in a mating pair of Chaser dragonflies.

Large White butterfly Pieris brassicae.

Orange-tip butterfly Anthocaris cardamines. This photograph is from the same sequence shown on page 68

Brimstone butterfly *Gonepteryx rhamni*. This photograph is from the same sequence as shown on page 86.

Reconstructed take-off sequence in a Small Tortoiseshell butterfly *Aglais urticae*.

(left and right) Green-veined White butterfly Artogeia napi. These are from the same sequence and separated in time by 1/60th of a second. Within that brief period the wing 'eclipsed' the sun.

Brimstone butterfly *Gonepteryx rhamni.*

Reconstructed take-off sequence in a Green-veined White butterfly *Artogeia napi*.

(left and right) Marbled White butterfly Melanargia galathea.

Painted Lady butterfly Cynthia cardui.

Reconstructed take-off sequence in a Small White butterfly Artogeia rapae.

Small Tortoiseshell butterfly Aglais urticae.

Brimstone butterfly *Gonepteryx rhamni*.

(left and right) Green-veined White butterfly Artogeia napi.

Reconstructed take-off sequence in Four-spotted Chaser dragonfly *Libellula quadrimaculata*.

Brimstone butterfly *Gonepteryx rhamni*.

(left and right) Ladybird. Single images from the sequence shown on page 117

Brimstone butterfly *Gonepteryx rhamni*.

Meadow Brown butterfly *Maniola jurtina*.

Painted Lady butterfly Cynthia cardui.

Reconstructed take-off sequence in a Brimstone butterfly *Gonepteryx rhamni*

Dragonfly. Note the propeller-like twist in the wings.

Flight sequence in a mating pair of Small White butterflies Artogeia rapae.

Large White butterfly Pieris brassicae.

Reconstructed take-off sequence in a Painted Lady butterfly Cynthia cardui.

Brimstone butterfly Gonepteryx rhamni.

Reconstructed take-off sequence in a Peacock butterfly *Inachis io*.